Henry C. Lang

The Butterflies of Europe Described and Figured

Volume II. - Plates

Henry C. Lang

The Butterflies of Europe Described and Figured
Volume II. - Plates

ISBN/EAN: 9783744653558

Printed in Europe, USA, Canada, Australia, Japan

Cover: Foto ©berggeist007 / pixelio.de

More available books at **www.hansebooks.com**

Rhopalocera Europæ.

THE
BUTTERFLIES OF EUROPE.

Rhopalocera Europæ

DESCRIPTA ET DELINEATA.

THE

BUTTERFLIES OF EUROPE

DESCRIBED AND FIGURED.

BY

HENRY C. LANG, M.D., F.L.S., &c.,

MEMBER OF THE ENTOMOLOGICAL SOCIETY OF LONDON.

ILLUSTRATED WITH MORE THAN EIGHT HUNDRED COLOURED FIGURES, DRAWN, MOSTLY
FROM NATURE, UNDER THE DIRECTION OF THE AUTHOR.

VOLUME II.—PLATES.

LONDON:

L. REEVE & CO., 5, HENRIETTA STREET, COVENT GARDEN.

—

1884.

CONTENTS OF PLATES.

* Wrongly written Cordula at foot.

ALPHABETICAL INDEX OF SPECIES AND VARIETIES
FIGURED ON THE PLATES.

1. Papilio Podalirius, L. 2. Var. Feisthamelii, Dup.
3. P. Alexanor, Esp. 4. P. Machaon, L.

H. C. Lang direx.
Horace Knight chr. lith. ad nat.

1. *Papilio Hospiton*, *Géné* 2. *Thais Cerisyi*, *B.*
3. *Var. Caucasica*, *Ld.* 4. *Var. Deyrollei*, *Oberth.*

1 Thais Polyxena, Schiff. 2. Var Cassandra, Mn.
3 T. Rumina, L. 4. Var: Medesicaste. Illig.
5. Var. Honoratii, B. 6 Doritis Apollinus, Höst.

H. Leny direct
Horace Knight the lith ad nat   West Newman & Co imp.

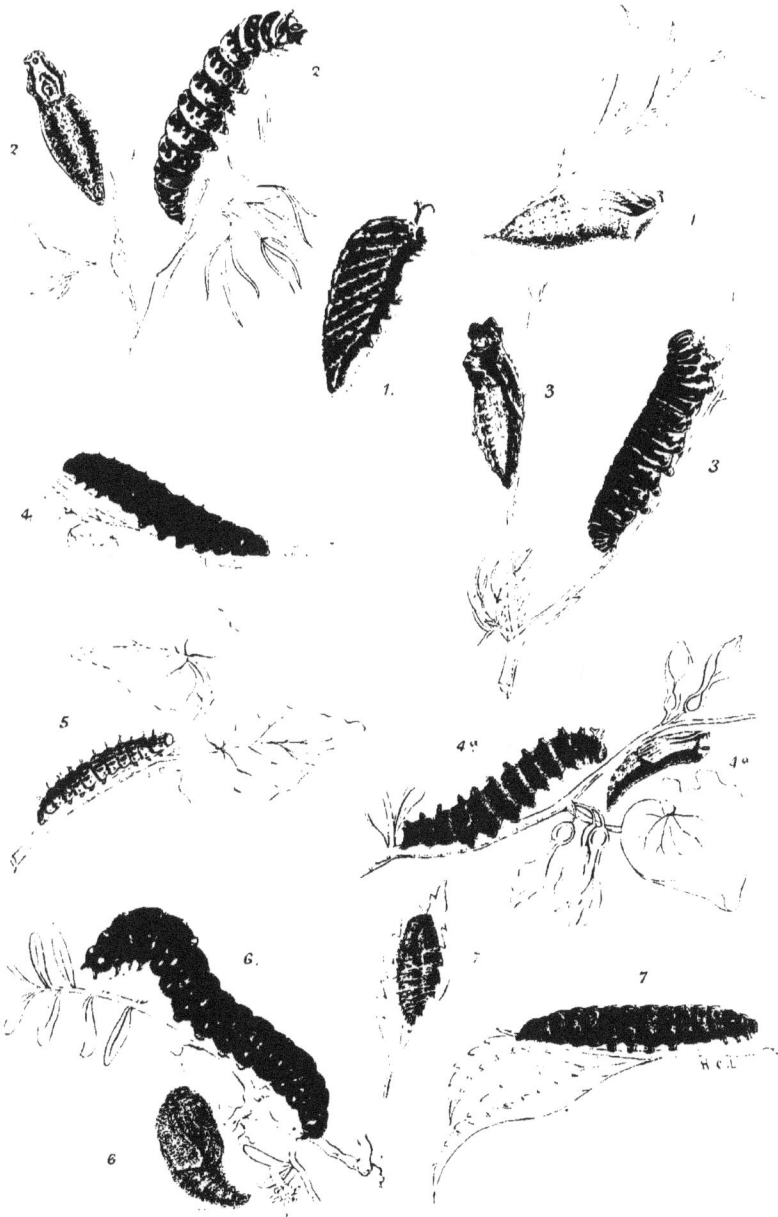

1. *Papilio Podalirius*. 2. *P. Alexanor*. 3. *P. Machaon*. 4. *P. Hospiton*.
4ª *Thais Cassandra*. 5. *T. Rumina*. 6. *Parnassius Apollo* 7. *P. Mnemosyne*

After Hubner H.C Lang direxi
Horace Knight chrom lith

L Reeve & Cº London.

West Newman & Cº imp

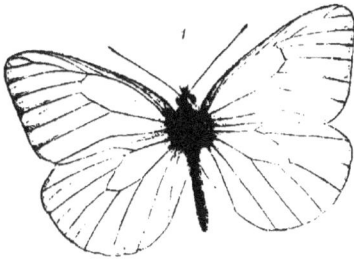

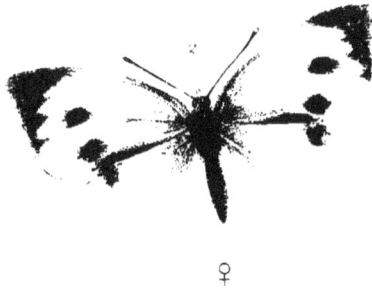

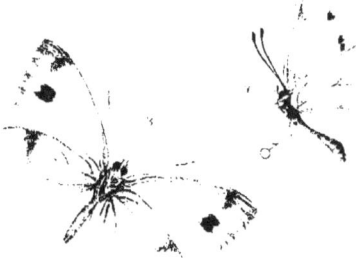

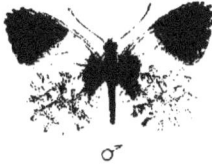

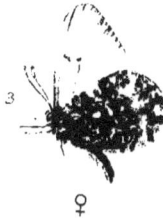

1. Euchloe Cardamines, L. 2. E. Gruneri, H.S.
3. E. Damone Feisth. 4. E. Euphenoides, Stg

1. *Euchlöe Pyrothöe. Ev.*   2. *Zegris Eupheme , Esp.*
3. *Zegris Menestho, Mén.*   4. *Leucophasia Sinapis,L.*
5. *Var. Diniensis, B*   6. *L. Duponcheli, Stgr.*

H.C. Lang direx!
H. Knight chr. lith ad nat.

Wert Newman & C° imp

L. Reeve & C? London.

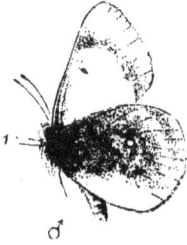

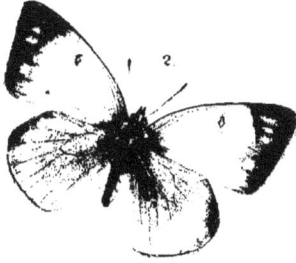

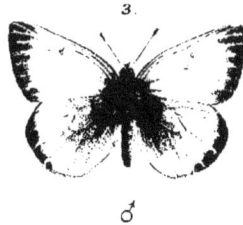

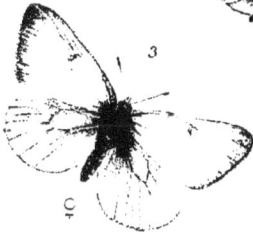

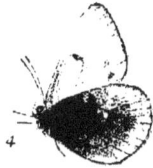

1. *Colias Palæno, L.*   2. *Var. ♀ Werdandi, H.S.*
3. *C. Pelidne, B.*   4. *C. Nastes, B.*

H.C.Lang direx*
H. Knight chr. lith ad nat.   West Newman & C.º imp

L. Reeve & C.º London.

1. *Colias Werdandi Zet*　　2. *C. Phicomone, Esp.*
3. *C. Hyale, L.*　4. *C. Erate, Esp.*　5. *Var. ♀ Pallida Stgr.*

H.C. Lang direx.ᵗ
H. Knight chr lith ad nat　　　　　　　　　　　West Newman & Cᵒ imp
　　　　　　　　L Reeve & Cᵒ London

1 Colias Edusa, F. 2 Var ♀ Helice. Hub. 3. C. Heldreichi, Stgr.
4. Gonepteryx Rhamni, L. 5. G. Cleopatra, L.

R.C.Lang dirext
H.Knight chr lith ad nat   L. Reeve & Cº London.   West Newman & Cº imp.

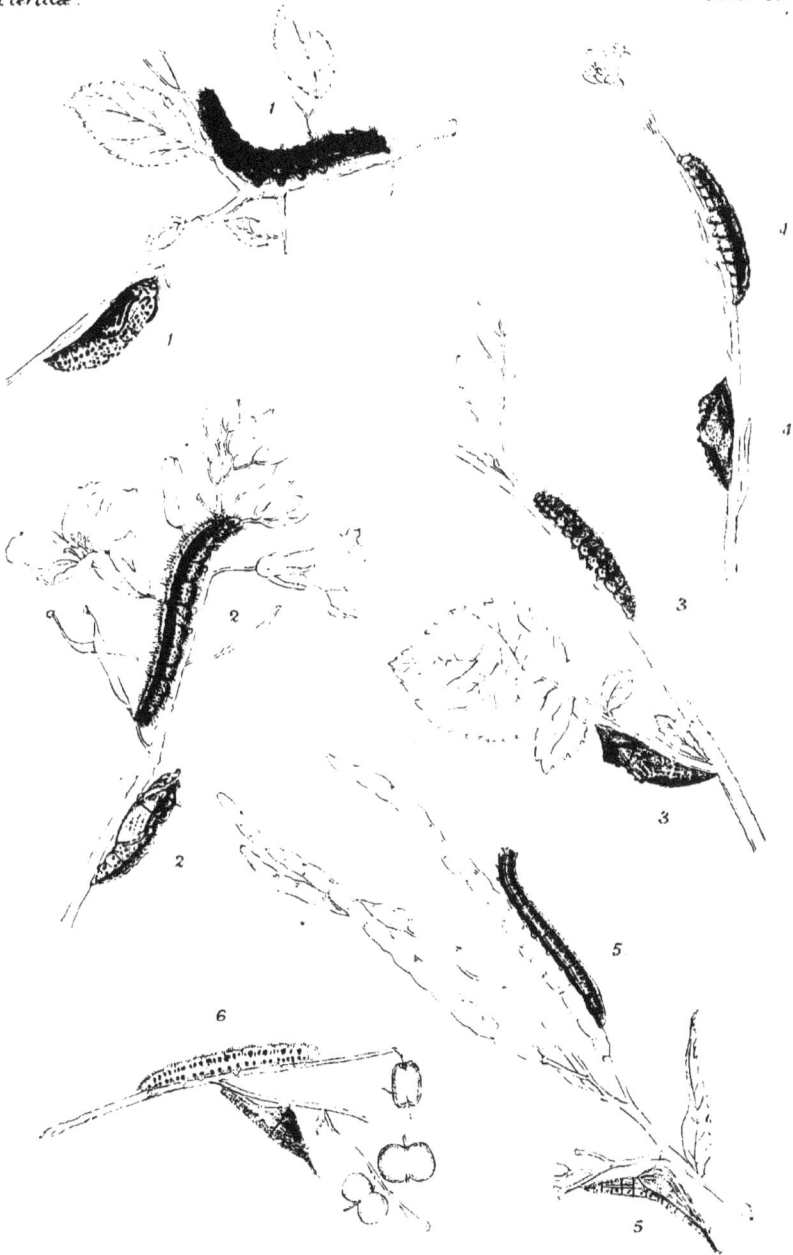

1. *Aporia Cratægi* 2. *Pieris Brassicæ.* 3. *P. Rapæ*
4. *P. Napi.* 5. *Euchloë Cardamines* 6. *E. Euphenoides.*

1. *Leucophasia Sinapis*. 2. *Colias Hyale*. 3. *C. Edusa*.
4. *Gonepteryx Rhamni*. 5. *G. Cleopatra*.

Arthur Hubner · H C Laing dirext
Horace Knight chrom. lith.

L. Reeve & Cº London.

West Newman & Cº imp

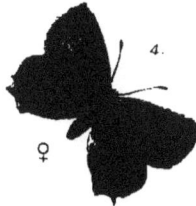

1. Thecla Pruni, L.   2. T. Quercûs. L.   3. T. Rubi. L.   4 Læosopis Roboris.
5 Thestor Ballus. F.   6. T. Nogellii. H-S.   7. T. Callimachus, Ev.

E. Lueng direx.ᵗ
J. Knight Jnr. lith ad nat

L. Reeve & Cᵒ London.

West Newman & Cᵒ ..

1 Polyommalus Virgaureæ, L.   2 P. Ottomanus, Lef.
3 P. Thersamon, Esp.   4 P. Dispar, Haw.

H. J. Lang direxit
F. Knight Ar lith ad nat   L. Reeve & Cᵒ London   West Newman & Cᵒ imp

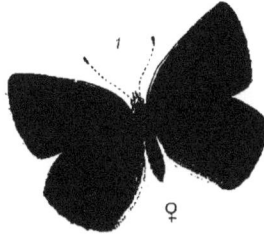

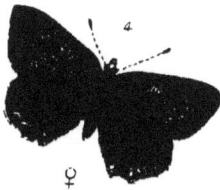

1. *Polyommatus Dispar, Var. Rutilus, Wernb.*   2 *P. Thetis, Klug*
3 *P. Ochimus, H.S.*   4 *P. Hippothöe, Lin*   5 *Var. Eurybia – O.*

H. L... del.   West Newman & Co. imp.
H. Knight m lith ad nat

L Reeve & Co Lowdon

1 *Polyommatus Alciphron Rott*   2 *P Gordius Sulz*   3. *P Dorilis, Hufn.*
4 *P Phlæas, Linn*   5 *Var Eleus F*   6 *P Helle W.V*

L. Reeve & C° London

1. Polyommatus *Phlæas*, var. *Schmidtii*. Gerh. 2 Lycæna *Bœtica*. L.
3. L. *Telicanus*, Lang. 4 L. *Balcanica*, Frr. 5. *Argiades* Pall.
6 L. *Fischeri*, Ev 7 L. *Trochilus*, Frr.

L. Reeve & Cᵒ London   West Newman & Cᵒ imp

1. *Lycæna Ægon,* s v   2 *L.Argus,* l   3. *L. Optilete,* Kn   4 *L. Zephyrus,* Eriv
5 *Var Hesperica,* Rbr   6 *L. Pylaon,* Fisch   7 *L. Pavius,* Ev.

H C Lang direx'
Horace Knight chrom lith ad nat   L. Reeve & C? London.   West.Newman & C? imp

1. L. Orion, Pall 2 L. Baton, Berg 3 L. Lysimon. Hb. 4. L. Rhymnus, Ev.
5 L. Psylorita Fr. 6. L. Pheretes Hb 7 L. Orbitulus, Prun. 8 var. Aquilo, B.
9. L. Astrarche. Bgs. 10. Var. Artaxerxes, Fab.

H.C. Lons direx.
Horace Knight chrom. lith. ad nat.
L. Reeve & Co London.
West Newman & Co. imp.

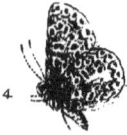

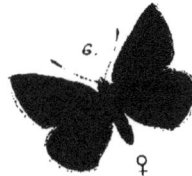

1. L. Anteros, Frr 2. L. Eros, O. 3. Var. Eroides, Friv.
4. L. Icarus, Rott. 5. Var. Icarinus, Scriba. 6. L. Eumedon Esp.

H. C. Long direx.
Horace Knight delin. lith. ad nat.   Vest Newman & Co. imp.
                        L. Reeve & Co. London.

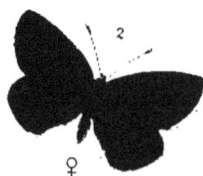

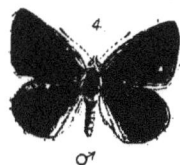

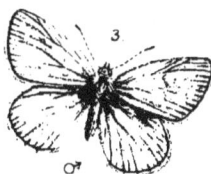

1. *Lycæna. Idas Rbr* 2. *L.Amanda Schn.* 3. *L.F.scheri, Hüb.*
4. *L. Bellargus, Rott.* 5. *Var.♀ Ceronus, Esp.* 6. *L.Corydon, Poda.*
7. *Var.♀ Syngrapha, Kef.* 8. *Var. Albicans, H.S.*

H.C.Lang, direxit
H Knight chr lith.ad.nat.

West,Newman & C⁰ imp

L.Reeve & C⁰ London.

1. *Thecla Betulæ*. 2. *T. Spini*. 3. *T. Ilicis*.
4. *T. Pruni*. 5. *T. Quercûs*. 6. *T. Rubi*.

After Hubner H.C Lang direct
.u we Hughs chrom. lith.

L. Reeve & Cº London.

West Newman & Cº imp

1. *Polyommatus Virgaureæ* 2. *P. Dispar v. Rutilus*
3. *P. Phlœas.* 4. *P. Amphidamas*
5. *Lycæna Bœtica*

J.C. Lang direx. after Hübner & Millière
Horace Knight chromo lith.

L. Reeve & C.º London

West Newman & C.º imp

1. *Lycæna Telicanus.* 2. *L. Argus.* 3 *L. Damon.*
4. *L. Cyllarus.* 5. *L. Melanops.*

H.C. Lang direx* after Hübner & Millière
Horace Knight chromo lith.

West Newman & C.° imp

L. Reeve & C.° London.

1. *Lycæna Hylas*, Esp. 2. *Var.Nivescens* Kef. 3. *L.Meleager*,Esp.
4. *L. Admetus*, Esp. 5. *L. Dolus*, Hüb. 6. *L. Damon*, Schiff.
7. *L. Donzelii*, B.

H.C. Lang dirext
R.Knight lith del. ad nat.

L. Reeve & C° London.

1. *Lycæna Argiolus* L. 2. *L. Sebrus*, B. 3. *L. Minima* Faessl.
4. Var *Lorquini*, H-S. 5. *L. Semiargus*, Rott 6. *L. Cælestina* Ev.
7. *L. Cyllarus*, Rott

H.G. Lang direx?
R Knight ad lith ad nat

West Newman & Co...

L. Reeve & Co L. m.a. u

1 *Lycaena Melanops*, B. 2.*L..Iolas*, O. 3.*L..Aicon*, F.
4 *L Euphemus*, Hüb. 5.*L.Arion*, L. 6.*L.Arcas*, Rott.

H.C.Lang dirext
B Knight lith ad nat

L.Reeve & C⁰ London.

W. Newman & C⁰ imp

1 *Nemeobius Lucina*, L.
2. *Libythea Celtis*, Fab.

1 Charaxes Jasius L   2 Apatura Iris, L
3. Var Iole. Schiff

Plate XXXV

1 Apatura Ilia Schiff   2 Var Clytie Schiff.
3 Var Metis Frr   4 Var Bunea , H S

1. *Charaxes Jasius.* 2 *Apatura Iris.* 3. *A. Ilia.*

H.C. Langdon? after Hubner
Horace Knight chromo lith.

L. Reeve & Cᵒ London

West Newman & Cᵒ imp

1. Limenitis Sibilla L. 2. Neptis Lucilla Fab.
3. N. Aceris Lepechin. 4. Vanessa Levana Lin.

H. G. Lang sc.
R. Künstler lith ad nat.
Ver. Nürmann b.

Plate XXXIX

1 Vanessa Xanthomelas, W.V. 2 V. Vau - album, W.V.
3. V Urticæ, L 4. Var Turcica, Staud. 5. Var. Ichnusa, Bon

W.C.Long dirext
R.Knight lith ad nat

L.Reeve & Cº London

Westfleuman & Cº imp

1. *Vanessa Io. Linn.* 2. *V. Antiopa. Linn.*
3. *V. Atalanta. Linn.*

1 *Vanessa Callirchöe Fab*   2. *V Cardui, Linn*
3 *Melitæa Cynthia, Hüb.*   4. *M Iduna Dalm.*

L. Reeve & Cº London

1 Melitæa Maturna, Linn. 2 M. Aurinia, Rott.
3. Var. Merope, Prun. 4 Var Provincialis, Boisd.
5. Var. Desfontainii Godt. 6. M. Bœtica, Rbr.

1 *Melitæa Cinxia*, Linn. 2 *M. Arduinna*, Esp. 3 *M. Phœbe*, Knoch.
4 Var. *Æthereα*, Ev. 5 *M. Trivia*, Schiff.
5 *M. Didyma*, O. 7 Var *Neerα* F. de W.

L. Reeve & C.º London.

1 *Melitæa Dictynna* Esp. 2 *M. Dewne* Hub. 3 *M. Athalia* Ros.
4. *M. Aurelia* Nick.   5. *M. Parthenie* Brh. 6 *M. Asteria*, Pre.

H.C. Lang *direx*.
H Knight chr. lith ad nat.

L. Reeve &   & Co.

1. *Argynnis Aphirape.* Hüb 2 *Var. Ossianus* Hbst 3 *A. Selenis*, Fv.
4. *A. Selene*, Schiff. 5 *A Euphrosyne*, L. 6 *Var. Fingal*, Hbst

H.C Lang direx'
H. Knight chr lith ad nat.

W. Newman & C° imp

L. Reeve & C° London

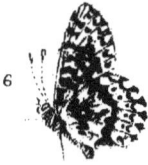

1 Argynnis Pales, Schiff. 2 Var. ♀ Napæa, Hüb 3 Var. Lapponica, Staud
4 Var. Arsilache, Esp. 5. A. Chariclea, Schn. 6. A. Polaris. B. 7. A. Freija, Thnb.

F. C. Knight dirext.
H. Knight lith. et dirext.   West Newman & C° imp.

L. Reeve & C° London.

1 *Argynnis Dia Linn.* 2. *A. Amathusia*, *Esp.* 3. *A. Frigga*, *Thnb.*
4. *A. Thore*. *Hüb.* 5. *A. Daphne*, *Schiff.* 6. *A. Ino. Esp.*

F. C. Lucas direx.t
P. T. Lucas del. et lith. ad nat.

Wm. Newman & C.° im.t

J. Roux & C.° impr.m

♀

♂

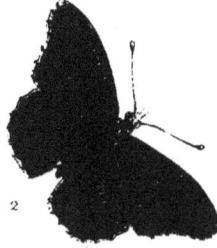

Plate L

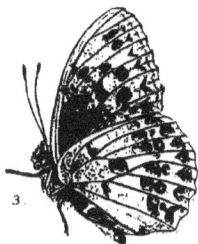

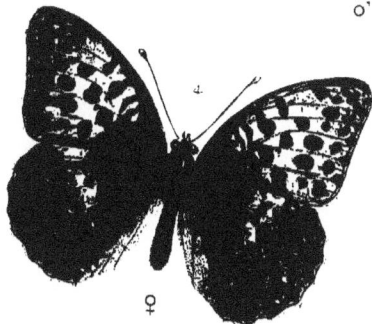

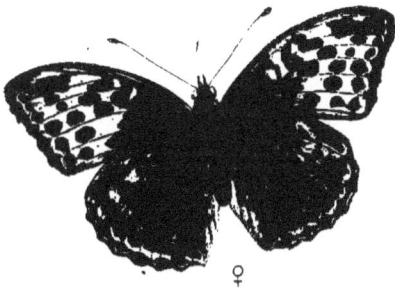

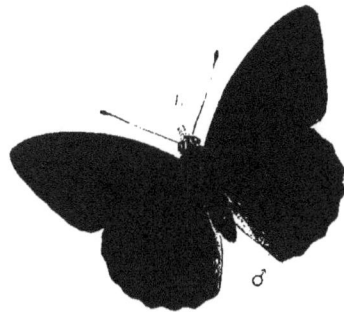

1 *Argynnis Paphia*, L   2. Var ♀ *Valezina*, Esp.
3. A *Pandora*, Schff.

F.C Lowe direx*
H. Knight de lith ad nat   West Newman & C°° imp

L Reeve & C° London.

After Hubner H C Lang direx*
Horace Knight chromo lith

1. *Limenitis Populi.* 2. *L Camilla.* 3 *Vanessa Levana.*
4 *Vanessa Antiopa.* 5 *Melitæa Cynthia.* 6. *Argynnis Amathusia.*
7. *A. Niobe.*

L. Reeve & Cº London.

West Newman & Cº imp

Danais Chrysippus, Linn.

After Hubner. H G Long direx.
Horace Knight chromʳ lith

West Newman & Cⁿ imp

L. Reeve & Cⁿ London.

1. *Melanargia Galatea* L. 2. Var ♀ *Leucomelas,* Esp
3. *Var Procida,* Hbst. 4. *& Lachesis.* Hub.
5. *M. Larissa var Herta* Hub

H.G.Knaggs direx!
H. Knight lith ad nat   West Newman, sc
J. Reeve & C⁰ London

1　Melanargia Larissa. Hub (form typ.) 2 M. Japygia Cyr
　　　　　　3. M. Pherusa, Boisd.

H C Lang direx.
H Knight ch lith. ad nat.

West Newman & Cᵒ imp.

L Reeve & Cᵒ London

1. Melanargia Syllius, Höst. 2 M. Arge, Sulz.
3. M. Ines, Hffsgg.

H.C Lang direct
H.Knight ch. lith ad nat

West Newman & Cº imp

L. Reeve & Cº London

1 *Erebia Epiphron,* L.   2 *Var. Cassiope Fab.* 3 *E. Melampus, Fuessl.*
4. *E. Eriphyle, Frr.*   5 *E. Arete Fab.*   6 *E. Mnestra Hub*
7. *E. Pharte Hub*   8. *E. Œme , Hub.*

J. Berre & C° London

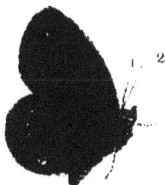

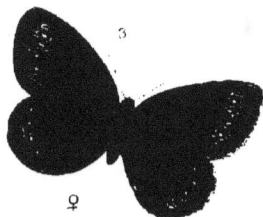

1. *Erebia Manto* Esp   2. *E. Ceto* Hub.
3. *E. Medusa* F.

H C Lang direx
H Knight del lith ad nat

West Newman & Co imp

Reeve & C° London.

H C Lang direx'
H Knight chr lith on val

West Newman & C° imp.

Reeve & C° London

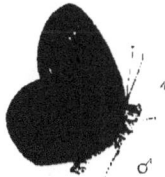

1. Erebia Stygne ♂   2. E. Nerine ♀♂.
3. Var. Reichlini, H-S.   4. E. Evias ♂♀

Erebia Melas, Hbst. 2. E. Glacialis Esp. 3. Var. Alecto Hüb
4. E. Scipio, Boisd. 5. E. Epistygne ; Hüb

H.L. long. pinx.
R. hosebirsht lith. ad nat.   West Newman & Co. imp.
                    L. Reeve & Co. London

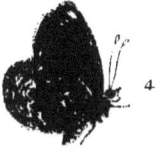

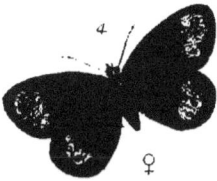

1. Erebia Afra, Esp. 2. E. Lappona. Esp. 3. E. Tyndarus. Esp.
4. E. Gorge. Esp.   5. E. Goante, Esp

H. C. Lang direx.
H. Knight imp. lith. ad. nat.   West Newman & C.º imp.

L. Reeve & C.º London.

1. Erebia Pronoë, Esp   2. Var. Pithio, Hüb.
3. E. Neoridas, Boisd.   4. E. Zapateri, Oberth.
5. E. Æthiops, Esp. (Medea Hüb.) 6. E. Ligea, L.

1. Erebia Euryale, Esp. 2. E. Embla, Thnb. 3. E. Disa, Thnb.
4. Œneis Jutta, Hüb. 5. Œ. Norna, Thnb.

H C Lang dirext
H. Knight ch. lith ad. nat.   West Newman & Cᵒ imp

L. Reeve & Cᵒ London.

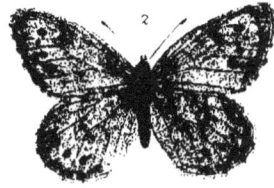

1. Œneis Aello, Hub.   2. Œ. Tarpeia, Pall.
3. Œ. Bore, Thnb.   4. Satyrus Alcyone, Schiff.

H. C. Lang direx.ᵗ
H. Knight chr lith ad nat   West Newman & Cᵒ imp

L. Reeve & Cᵒ, London

1. Satyrus Hermione, L.   2 S. Circe Fab.
3. S. Briseis, L.

H C Lang direx*
H. Knight ch. lith ad nat

L. Reeve & C°., London

West Newman & C° imp

1 S Anthe, O.   2. Var. Hanifa, Nord. 3. S. Autonöe, Esp.
4. S. Semele, L.   5 Var Aristæus, Bon.

H.C Lang direx.ᵗ
H.Knight ch lith.ad nat.

L. Reeve & Cº. London.

West Newman & Cº imp

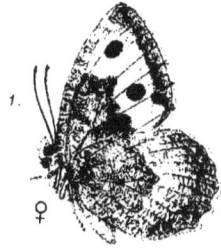

1 *Satyrus Græca*, Stgr. 2. *S Amalthea*. Friv.
3♂.Hippolyte, Esp.

H.C.Lang direx.ᵗ
H Knight del.. lith ad nat.

West. Newman & Cᵒ imp

L.Reeve & Cᵒ, London.

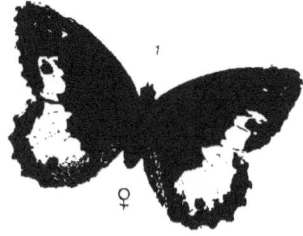

1. *Satyrus Neomiris. Godt*   2. *S. Arethusa Esp*
3. *Var Dentata Stgr.*   4. *S. Statilinus Hufn.*

H.C.Lang direx.t
H. Knight ch. lith ad nat.   West Newman & C.º imp

L. Reeve & Cº London.

1. S. Statilinus var. Allionia, F   *2. Var Fatua, Frr.*   *3. S Fidia, L.*
       4. S. Dryas, Sc   *5. S Cordula, F.*

H.C Lang direx.
H Knight ih lith ad nat   West Newman & Cº imp

L. Reeve & Cº London.

1 Satyrus Cordula, F. 2 Pararge Roxelana Cr. 3. P. Clymene Esp.
4 P. Hiera, F. 5 P. Megæra, L. 6. Var Tiadius, Bon.

H.C Lang direx.
H Knight lith ad nat.   West Newman & Co

L Reeve & Co London

1. *Pararge Mœra*, L. 2. *P. Egeria*, L. 3. *Var. Egerides*, Stgr.
4. *P. Achine*, Sc. 5. *Epinephele Nurag*, Ghil.

H.C. Lung direx?
H.Knight lith ad nat.

Vinc. Brooks, Day & C° imp.

L. Reeve & C° London

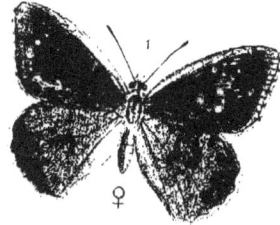

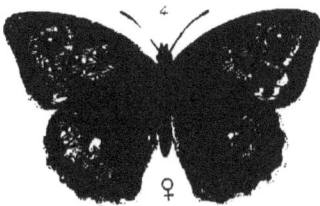

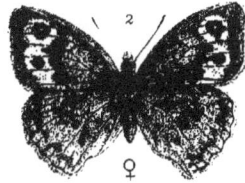

1 Epinephele Narica, Hüb   2. E Lycaon, Rott
3. E Janira. L. 4 Var Hispulla Hüb.

H.C Long del et
H.Knight ch lith ad nat.   West Newman & Cº imp
                          L. Reeve & Cº London

1. Epinephele Ida Esp   2. E. Tithonus, L.
3. E. Pasiphae, Esp.   4. E. Hyperanthus, L.
5. Ab. Arete, Müll.   6. Cœnonympha Œdipus F.

H C Lang direx⁺
H. Knight ch lith ad nat   West Newman & Cᵒ imp

L. Reeve & Cᵒ London

1. Cœnonympha Hero, Linn. 2 C. Leander, Esp 3. C. Iphis, Schiff.
4. C Arcania, Linn. 5 var Darwiniana, Star. 6. C. Satyrion, Esp.
7. C. Corinna, Hüb. 8 C. Amaryllis, Cr 9. C. Thyrsis, Frr.

H.C Lang direx^t
H. Knight ch lith ad nat
                                                                    West Newman & Co imp
                              L. Reeve & Co London

1. *Melanargia Galatæa.* 2 *Erebia Medusa.* 3. *E. Ligea.*
4. *Satyrus Circe.* 5, *S. Dryas.* 6. *Pararga Achine.* 7. *Epinephele Lycaon.*
8. *Cœnonympha Arcanius.* 9. *C. Iphis.*

after Hubner. H C Lang direx'
Horace Knight chrom. lith.

West Newman & C° imp

L. Reeve & C° London.

H.C.Lang dirext
H.Knight ch lith ad nat.   L. Reeve & Cº. London   West.Newman & Cº imp

1. *Cænonympha Dorus Esp.* 2 *C.Tiphon, Rott.* 3. *Var. Philoxenus, Esp.*
4. *Var. Isis, Thnb.*   5. *C. Pamphilus, Linn.*   6 *Var. Lyllus, Esp.*
7. *Triphysa Phryne. Pall.*
SUPPLEMENTARY FIGURE.   8 *Argynnis Improba, Butl.*

Plate LXXVIII

1. *Spilothyrus Alceæ*, Esp. 2. *S. Altheæ*, Hub 3 *S. Lavateræ*, Esp.
4. *Syrichthus Proto*, Esp 5. *S. Tessellum*, Hub. 6. *S Cribrellum*, Ev.
7 *S. Cynaræ*, Rbr. 8. *S. Sidæ* Esp 9 *S. Carthami*, Hub.

H.C Lang direx.t
H.Knight ch.lith ad nat.

West Newman & Cº imp.

L.Reeve & Cº, London

1. Syrichthus Alveus, Hüb.   2. Var. Cirsii, Rbr.   3 Var. Carlinæ, Rbr.
4. S. Serratulæ Rbr.   5 Var. Cœcus, Frr.   6. S. Cacaliæ, Rbr
7. S. Andromedæ, Wallgr. 8 Centaureæ Rbr.

H.C.Lang dirext
H Knight ch. lith ad nat.   West.Newman & Cº imp

L.Reeve & Cº London.

After Hubner: H.C.Lang direxᵗ
Horace Knight chrom lith.   West Newman & Cᵒ imp

1. *Spilothryrus Alceæ.*   2. *Syrichthus Malvæ.*
3. *Nisoniades .Tages.*   4. *Hesperia Linea.*
5. *Carterocephalus   Palæmon.*

L.Reeve & Cᵒ London.

1. *Syrichthus Malvæ*, Linn.   2. *ab Taras*, Meig.   3. *S. Phlomidis*, H.S.
4. *S. Orbifer*, Hüb.   5. *S. Sao*, Hüb.   6. *Var. Therapne*, Rbr.
7. *Nisoniades Tages*, Linn.   8. *N. Marloyi*, B.   9. *Hesperia Thaumas*, Hufn.
          10. *H. Lineola*, O.   11. *H. Actæon*, Esp.

H.C. Lang direx.
H.Knight di lith ad nat   West Newman & Cº imp.

L. Reeve & Cº, London

1. *Hesperia Sylvanus,* Esp   2. *H. Comma,* Linn.
3. *H. Nostrodamus,* Fab.   4. *Cyclopides Morpheus,* Pall.
5. *Carterocephalus Palæmon,* Pall. 6. *C. Sylvius,* Knoch.
SUPLEMENTARY FIGURE.
7. *Lycæna Lycidas,* Trapp.

H.C.Lang dirext
H.Knight ch lith ad nat.

West Newman & Cº imp

L. Reeve & Cº London.